DATE DUE

MAY 2 1983		
MAY 3 1 1983		
JUN 2 8 1983		
AUG 2 1983		
AUG 2 6 1983		
SEP 1 3 1983		
NOV 1 1983		
DEC 3 0 1983		
JUL 1 3 1984		
FEB 2 5 1985		

From Cotton to Pants

From Cotton
to Pants

Ali Mitgutsch

 Carolrhoda Books, Inc., Minneapolis

First published in the United States of America 1981 by
Carolrhoda Books, Inc. All English language rights reserved.

Original edition © 1977 by Sellier Verlag GmbH, Eching bei München,
West Germany, under the title VON DER BAUMWOLLE ZUR HOSE.
Revised English text © 1981 by Carolrhoda Books, Inc.
Illustrations © 1977 by Sellier Verlag GmbH.

Manufactured in the United States of America

LIBRARY OF CONGRESS CATALOGING IN PUBLICATION DATA

Mitgutsch, Ali.
From cotton to pants.

(A Carolrhoda start to finish book)
Edition for 1977 published under title: Von der Baumwolle zur Hose.
SUMMARY: Traces the journey of cotton from the
plants, through the cotton gin and the spinning mill
where it is made into thread, to the loom where it is
woven into cloth, and finally to the clothing factory
where it is sewn into pants.

1. Cotton manufacture—Juvenile literature. [1. Cotton]
I. Title.

TS1576.M5713   1981   677'.21   80-29552
ISBN 0-87614-150-5

1 2 3 4 5 6 7 8 9 10 86 85 84 83 82 81

From Cotton to Pants

Cotton grows on plants in clusters called **bolls**.
When the cotton is ripe,
many bolls are picked off each plant
by workers or by machines.
Each boll contains seeds and **cotton fiber**
and usually some dirt.

Before the cotton fiber can be spun into thread,
it must be cleaned.
The fiber is put into a large machine
called a **cotton gin**.
The cotton gin separates the seeds
and other things from the cotton fiber.
Then the clean fiber is taken to a **spinning mill**
where machines twist it into thread.

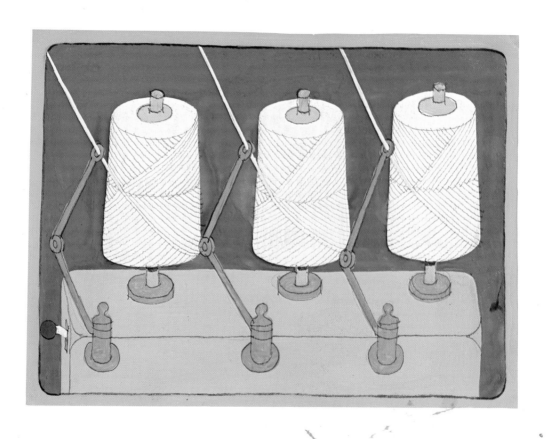

The cotton thread can now be woven into cloth.
Many threads are stretched side by side
on a machine called a **loom**.
Then other threads are woven over and under
the side-by-side threads.
The thread is woven together very tightly
to make strong, smooth cloth.
If threads of different colors are used,
a design can be woven into the cloth.
Plain cloth can be dyed.
Or a design can be stamped on it.

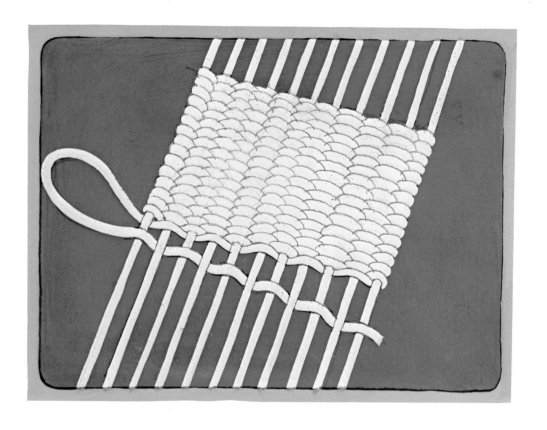

The finished cloth is wound
around heavy pieces of cardboard.
These rolls are called **bolts**.
Many bolts are loaded onto a truck
and taken to a clothing factory.

At the clothing factory
there are patterns for many kinds of clothing.
The patterns come in different sizes
so there will be clothing to fit everyone.

The **cutter** rolls out a bolt of blue cloth.

He lays a pattern on top of it.

The pattern shows him exactly where to cut.

Can you guess what he is cutting out?

Then the pieces are carefully sewn together.

The **seamstress** follows the pattern exactly.

She sews on pockets, patches, and belt loops.

The finished pants will go to a store to be sold.

These children are playing in pants
made from cotton cloth.
Cotton pants are strong
and easy to take care of.
And they are soft and comfortable too!

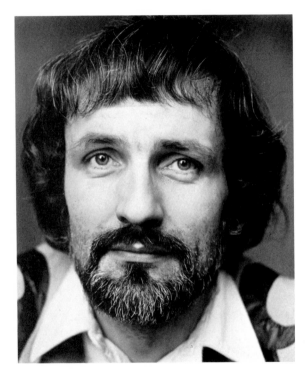

Ali
Mitgutsch

ALI MITGUTSCH is one of Germany's best-known children's book illustrators. He is a devoted world traveler, and many of his book ideas have taken shape during his travels. Perhaps this is why they have such international appeal. Mr. Mitgutsch's books have been published in 22 countries and are enjoyed by thousands of readers around the world.

Ali Mitgutsch lives with his wife and three children in Schwabing, the artists' quarter in Munich. The Mitgutsch family also enjoys spending time on their farm in the Bavarian countryside.

THE CAROLRHODA

START

TO FINISH BOOKS